Meeting People via WiFi and Bluetooth

Meeting People via WiFi and Bluetooth

Joshua Schroeder

Contributing Editor
Henry Dalziel

ELSEVIER

AMSTERDAM • BOSTON • HEIDELBERG • LONDON
NEW YORK • OXFORD • PARIS • SAN DIEGO
SAN FRANCISCO • SINGAPORE • SYDNEY • TOKYO
Syngress is an imprint of Elsevier

Syngress is an imprint of Elsevier
225 Wyman Street, Waltham, MA 02451, USA

Notices
Knowledge and best practice in this field are constantly changing. As new research and experience broaden our understanding, changes in research methods or professional practices, may become necessary.

Practitioners and researchers must always rely on their own experience and knowledge in evaluating and using any information or methods described herein. In using such information or methods they should be mindful of their own safety and the safety of others, including parties for whom they have a professional responsibility.

To the fullest extent of the law, neither the Publisher nor the authors, contributors, or editors, assume any liability for any injury and/or damage to persons or property as a matter of products liability, negligence or otherwise, or from any use or operation of any methods, products, instructions, or ideas contained in the material herein.

ISBN: 978-0-12-804721-7

British Library Cataloguing-in-Publication Data
A catalogue record for this book is available from the British Library

Library of Congress Cataloging-in-Publication Data
A catalog record for this book is available from the Library of Congress

For Information on all Syngress publications
visit our website at http://store.elsevier.com/Syngress

CONTENTS

ABOUT THE AUTHORS

Henry Dalziel is a serial education entrepreneur, founder of Concise Ac Ltd, online cybersecurity blogger and e-book author. He writes for the Concise-Courses.com blog and has developed numerous cybersecurity continuing education courses and books. Concise Ac Ltd develops and distributes continuing education content (books and courses) for cyber security professionals seeking skill enhancement and career advancement. The company was recently accepted onto the UK Trade & Investment's (UKTI) Global Entrepreneur Program (GEP).

The author, **Joshua Schroeder**, originally from North Carolina, first started learning about computers from his dad, Donald Schroeder, a Senior Service Engineer. Most of his work consisted of assistance in applications, service, training, and sales for several different companies supporting the manufacturing efforts of the South East United States. Many nights he would bring home computers and other bits of technology for Joshua to play on. His skills and interest in the field deepened when his mother, Michelle Schroeder, who homeschooled him K-12, decided to start taking him weekly to the Hispanic Learning Center (today known as the International Center for Community Development) where he helped design, run, and fix their computer network to help students learn reading, writing, and math in their after-school programs. This foundation of skills propelled him to be successful in Information Technology and Security, and for their help and guidance he is very thankful.

In college, he graduated with a Masters in IT with a Concentration in Security and Privacy from the University of NC at Charlotte (UNC Charlotte) where he also is known for helping found the 49th Security Division—a security club that still helps to instill interest in security and other pieces of technology for students at the university. Since college, his research and work include presentations at ShmooCon and SkyDogCon titled "CCTV: Setup Attack Vectors and Laws," Carolina Con where he presented "Spam, Phish and Other Things Good to Eat," and "Burp Suite: A Comprehensive Web Pen Testing" as well as participation in the security group NoVA Hackers in Virginia.

He currently works doing incident response in Northern Virginia, however this research is different from the responsibilities of his job. Many of the tricks and tips included in this book would not have been possible without the training and guidance of the team that helps run the Wireless Village at DEFCON, and for their dedication year after year, he would like to express his thanks.

OVERVIEW

This chapter will contain an overview of how to track people using Wireless 802.11 radio frequencies (Wi-Fi) and Bluetooth 802.15 radio frequencies. The content contained here came from research and materials originally presented at Defcon Wireless CTF Village in August 2015 entitled "Meeting People Via Wi-Fi."

We will go over the hardware and software needed in order to do this tracking, how to use these particular tools in order to do attribution, and tips for protecting yourself from being attributed via those signals.

CHAPTER 1

Overview: Scanning for Device Association

Overview

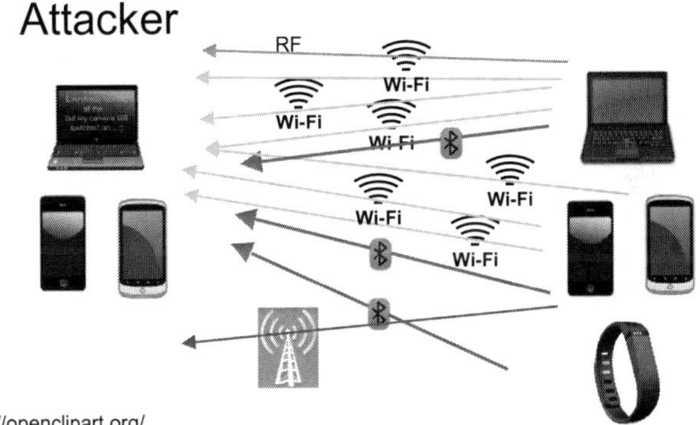

Source: https://openclipart.org/

Every day people use all kinds of devices that emit radio frequencies (RFs) that provide identifying information about them. Over the years, governments, private investigators, and radio enthusiasts have designed and created techniques to triangulate and identify individuals for the purpose of catching illegals, finding missing persons, shutting down radio interference, as well as intercepting communications.

The devices used every day to make common data communications allow us to connect to the Internet or internal networks in our home, work, or car. Many of these help us be more productive, safer, and stay in touch with those we know and love. However, with every connection we make, there exists the opportunity for something called RF signal leakage.

RF signal leakage occurs when someone other than the intended recipient receives a RF. This may be attacker related, such as running a scanner or interceptor, or accidental like when a cordless phone or radio picks up another cordless phone using the same channel. Either way this gives someone with nefarious intentions an opportunity to profile and track their target's daily movements and, in some cases, the possibility to intercept their communication.

RF signal leakage is not limited to audio transmissions, it can affect data and video too. If High-Definition Multimedia Interface (HDMI) cables or external keyboards used in a target's home or business consist of unshielded cables, then RF signal leakage could present a problem. An attacker could listen to and decode to keys typed or replay and decode images transmitted via the HDMI cable. This is why many keyboards, such as the Logitech, now encrypt their communications with encryption standards such as AES (Logitech, Inc., 2015). Even with encryption though, some of the underlying technology indicators like device type and distance from the sniffer or scanner could be determined, as these factors present themselves in order to establish the connection.

Another example, the smartphone, presents the opportunity for several vectors of profiling. Some, but not all, would include cell phone signal transmitting via Global System for Mobile (GSM) or Code division multiple access (CDMA), the probing for Wi-Fi hotspots to save on data rates or Bluetooth to connect to external peripherals. All of these different RF signals can essentially be used to identify a person in a given space.

This can be accomplished through the unique identifiers that tell the network, be it cell towers, Wi-FI, or point to point (P2P) Bluetooth, which allow the device to send and receive the communications. For example, each SIM card contains a unique identifier that tells the tower who you are so the provider can know what phone number and data allowance to give your device with a given number. The same is true with Bluetooth and Wireless communications, except those two technologies normally use Media Access Control (MAC) address.

All communication devices possess a Federal Communications Commission issued device MAC address. This identifier allows devices to communicate and prevents collisions and preserves the uniqueness of a device on a network. Bluetooth also has UUID codes that it displays during point-to-point association searches that are unique per device and could assist in profiling.

There may be additional IDs and signatures that can be associated but all Wi-Fi, Bluetooth, Cell Phone Radio Frequencies, or some other similar connections, it will have a MAC address in the form of

XX:XX:XX:XX:XX:XX, where the XX's will be replaced with a given unique MAC consisting of hex characters. The format for Wi-FI and Bluetooth consists of a format of XX:XX:XX:AA:BB:BB, where the first three hexadecimals are set aside for the device manufacturer who purchased the license to create the device, and the last three hexadecimals represent the ID that the manufacturer assigned to that device. We will show later how this can be used to create a comprehensive target profile. This allows an attacker to create a profile, or database key, every time they see communication from that device as well as giving them a format to lookup device type in online databases provided by the FCC and other network companies.

It is important to note (according to a recent article www.arstechnica.com) that iPhones with iOS 8.0 and above have started to spoof MAC addresses. Based on a variety of factors, the most notable of which is a reboot of the phone forces, this protection technique has to be triggered. Further research shows a MAC address from an iPhone incrementing its address by one hex digit via Wi-Fi probes when searching for an access point.

Also, when attempting to find some device types, the manufacturer code sometimes replied as Private or Unknown which means that they are either protected from disclosure by the FCC (in the interest of National Security), or may be protected under development laws for the interest of protecting patents or new technologies, as explained by John Abraham, the creator of the Android Bluetooth scanning tool BlueScan.

Before we continue, let's clarify that all of this profiling can be done by means coined under the term of "War Driving," a term first established by Peter Shipley to mean "to go out and search for open wireless LANS." The act of war driving is legal; the act of breaking encryption on networks is not. If the attacker were to decrypt encrypted commutations without permission (exceptions from permission given by the target, or a court-issued order), this would in most court cases be illegal because the attacker is then breaking the basic concepts contained within laws surrounding reasonable expectation of privacy. The concepts and techniques talked about later will focus on legal means for conducting surveillance without the need for consent or permission. That being said, because laws change and are different from place to place, if you have concerns please contact an attorney.

The two most common RF signals, Wi-Fi and Bluetooth, can tell us enough about a suspect that would allow us to see where they have been, who they may have communicated with and make inferences about their exact current locations and how many times they have visited the location. All of these communications allow an attacker to know something about the target that can be categorized as radio signal leakage; and pretty much any device has it to some extent.

Most times Wi-Fi communication is a result of an attempt to establish Internet connection, or transfer data to a base station (such as a home router or access point) connected to a backbone, which has ties with a physical location. Knowing the name of the base station's **S**ervice **S**et **ID**entifier, or SSID, allows for the ability for someone to lookup in a database the unique name of that base station and allows the attacker to know where the target has visited (*PC Magazine Encyclopedia*. 22 October 2015).

Additionally, point-to-point communications such as tethering when someone uses their phone for Internet access on a laptop or tablet presents the ability to determine other identifiers. For example, if the attacker saw that someone had connected to a hotspot named "Verizon Mifi 877624" they could infer that the target uses Verizon. This would allow them to stage additional attacks or potentially figure out who is making those communications based on physical activities present in the environment.

Bluetooth, on the other hand, normally shows itself through point-to-point connections; it has a lower range and can be used to help pinpoint a target or determine if they have been to a location before.

For example, you could put a Bluetooth scanner on cell phone plugged into the wall that is located near an entrance to a building. From the Bluetooth MAC address of all the users who pass through, you could find out what percent of employees have BlackBerrys, iPhones, or Android Devices. The attacker could also profile and log physical fitness devices that people wear themselves, such as a Fitbit or Garmin, since these gadgets transmit via Bluetooth to the phone in order to help understand what's going on with the target's body. Every device someone carries is just another data point that allows the attacker to determine something about them or the people who travel in the same group as them. If an attacker did this for days at a time, he could pair it with CCTV information and build a database of all users and their device profiles.

Additionally, this profiling could be used to determine who comes to work on time or when someone doesn't show up on a certain day. If an attacker created a database of all the law enforcement officers or security guards who go in and out of a given building, they could later keep a device on them that alerted them to the presence of a guard or officer when they came close by identifying their Fitbit or cell phone (that they keep on them at all times).

The problem is that we need these in order to go about our daily lives. You have to use it in order to connect to the Wi-Fi hotspot at work, home, or in the airport or coffee shop. You need Bluetooth in order to talk on the cell phone without having your hands leave the wheel. You can encrypt your connections, using a VPN or SSL encryption, but this doesn't stop the identifying vectors from having RF signal leakage.

What is the real value of collecting the MAC address of a device of a target? Well, this would be important information if the attacker were going to follow up that information with a spear phishing campaign or SMS attack on phone numbers associated with the organization. When you have the MAC address of a device you can determine answers to the questions such as: Is it an iPhone? Or an Android? Many times exploits are targeted at specific type of phone or device; knowing what a company employees use day to day could allow them to purchase the right exploits or develop an attack with a greater success rate.

Governments and Intelligence agencies have designed SCIFs, Sensitive Compartmented Information Facilities, to protect communications from leaving the secure facility. What about the naming schemes of the wireless at other parts of the complex? If the hotel on base says "Base1215-USAF-Wi-Fi," we can now figure out where that employee is staying and by other open-source intelligence (Google Maps, Microsoft Maps, Yahoo Maps, etc.) may be able to guess where they work. Research shows that many hotels, schools, and other places that have large numbers of users publicly post instructions for how to gain access to the network in PDFs or instructional websites online. These instructions are indexed by search engines, which then allows for easy correlation to a given location.

We need to learn what data leakage we have in order to understand how vulnerable our devices make us and learn how to protect ourselves a little bit better.

When it comes to Wi-Fi, every single Wi-Fi access point you've ever connected to is stored in an access list on your computer or cell phone or other device. Think of it as a web history for your network connections. The advantage to keeping this information is that it allows you to connect to a network without having to type in your password again or remember what the name of the Wi-Fi access point was for that particular location. The disadvantage of this is that they broadcast the SSID that you are trying to connect to.

For instance, we might have a hotel lobby that the target previously connected to but even if I wasn't in that location, my computer would say, "Hey, is hotelwifi1 lobby around?" And it would attempt to connect to those and tell everyone around that the target is looking for that particular Wi-Fi hotspot. As an attacker, we can pick these communications up using specific tools and gain attribution about the persons attempting to make those connections. These particular wireless hotspot names or SSIDs can give away information about where the target lives, works, or even the networks of friends who have connected to in the past.

A similar thing can also be done for Bluetooth. In discovery mode, when two Bluetooth pieces of technology are attempting to connect to each other, they will distribute the MAC address or discovery information, as well as other identifying features of the particular device. And even if they aren't in discovery mode, which is the set-up process for Bluetooth device, they are still attempting to connect and find other Bluetooth devices in the area that they previously connected to. This is good for the attacker, since they can just listen and don't have to alert the target that they are picking up the device. The attacker can also use something called "decibel" to guess the distance they are from the target's device. Decibel, when it pertains to RFs, is measured in negative numbers, unlike the decibel for audio which is positive numbers. The closer it is to zero, the closer the scanner is to the actual target device. An attacker could then use triangulation to locate the target.

There are three types of Bluetooth technology that are currently being used today in version 4.0 of the spec in everyday devices. Bluetooth in class 3 radios has a range of 1 m or 3 feet. Class 2 radios, which generally you will find in mobile devices, have a range of 10 m or 33 feet. And finally, Class 1 radios, the ones that are used in factories or other industrial areas in order to track shipments or other pieces of equipment, have a much longer range of about 100 m or 300 feet (Bluetooth SIG, Inc., 2015).

Class 3 Bluetooth is also known as Low Energy. Generally you see this in long-term Bluetooth sensors, such as on a fire hydrant or a fire extinguisher, or stored somewhere in a building where you are not going to be using it every single day but you want it to last for years, and it will have something like a 9-volt or a 5-volt battery that doesn't need a lot of charge in order to power it.

Class 2 has been referred to as "Classic." This is the one we think of today when we have our iPhones or Androids on and send pictures to each other using Bluetooth or communicate using a headset. This requires a lot more battery power and it will drain more quickly but it has a lot more features. We also see a lot of iPhones and Androids, and sometimes we see entertainment systems. If I was in an apartment complex where people have various devices that they are using in order to stream music, I'll sometimes pick up those devices and be able to share some features about them. It's not as useful as a Fitbit on iPhone, where I can track the person as they move around a particular geo-demographic, like a subway or their home, but it does give me some context and some information that I can use in order to figure some things out.

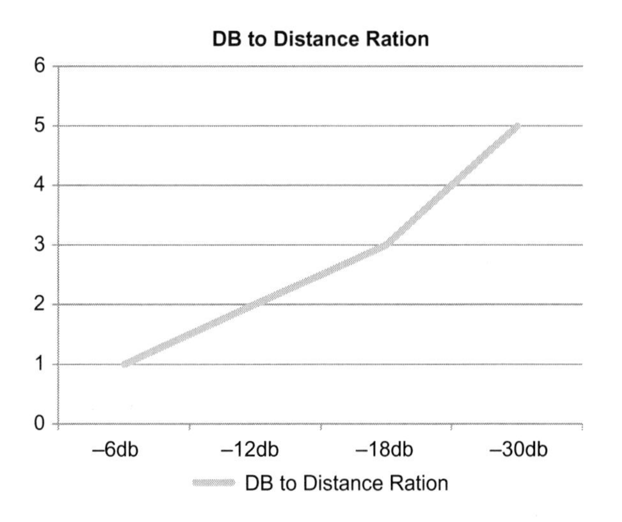

I mentioned that we use decibels in order to figure out how close we are. Notice that if we were to graph this relationship out, at zero we are very close to the device. And as the decibels become farther away from zero, the distance between the attacker and the device also increases.

Overview: Scanning for Device Association

100 meters

Class 3 radios – have a range of up to **1 meter** or 3 feet. Class 2 radios – most commonly found in mobile devices – have a range of **10 meters** of **33 feet**. Class 1 radios – used primarily in industrial use cases – have a range of **100 meters** or 300 feet.

Basics | Bluetooth Technology Website
www.**bluetooth**.com/Pages/Basics.aspx

This is helpful for knowing how close to the target the attacker is based on the technical limitations of the device. Other factors to consider are weather conditions where, for example, humidity and rain restricts signal ranges, as well as buildings and walls.

The goal of Bluetooth is honing—again, very close to the target—while Wi-Fi is for longer ranges. We have several different versions of Wi-Fi. There is A, B, G, and N, as well as AC, all having varying degrees of distance. All of these are in the form of the standard 802.11x, where x is the appending letter of the standard. If an attacker positioned in a busy mall to track their target, in most cases they would be able to find them relatively easily. Compare that to Bluetooth, where you would only need to be in the same room or same store as that particular individual.

Other factors to understand are that if you are trying to pick up AC signals with a B or G device, in most cases unless the device checks those other frequencies the attacker would miss those signals because they work on totally different frequencies. Also, each letter has different channels, for example, G has 1-11 in the United States and up to channel 14 in other countries. When an attacker is scanning the device, they can only listen on one channel at a time and the process of collecting information from all the channels is called channel hopping. To prevent missing anything, if possible use 4 cards on 1, 6, and 11 and 14 (if applicable). That way the card won't miss activity on channel 1 when it is hopping to a high channel like 7 or 8.

Hardware and Software Needed

Hardware Needed - Wifi

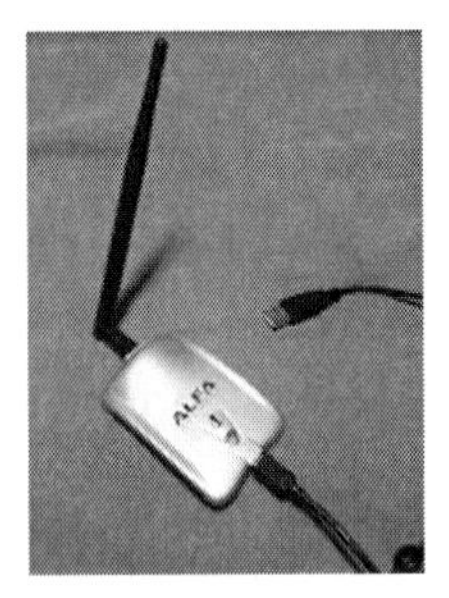

In this book, we are going to discuss a tool called the ALFA 802.11b/g RTL8187 chip set. This particular piece of equipment is really versatile and is used by a lot of pentesters because there is an imported version that works on Androids. I like it because it is relatively small; I can fit it in my pocket and I can be inconspicuous when I'm going about scanning. It also works in a promiscuous mode on Linux and Windows in order to allow me to pick up things there if I so choose. For example, if I was in a coffee shop I might be able to use it on targets using Android devices.

Hardware Needed - Wifi

Showing Today on Linux and Android + OTG Adapter

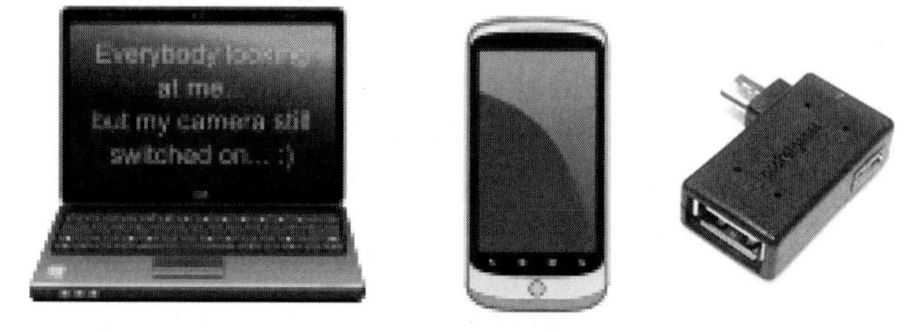

In my case study, we will use a Linux laptop, an Android, and an OTG adapter, which stands for "On The Go" adapter. What the OTG adapter allows us to do is plug the USB cable of the ALFA cord into the Android device. So you plug the OTG cable into the bottom where you would charge it and then you sequentially plug the actual ALFA cord into that and now you are able to use a particular Wi-Fi scanning tool on the Android using that external device. As far as software needed, we're going to use Wigle Wi-Fi for what's called war driving. This will allow us to access the database of previous war drives and pick up the access point.

Essentially, people travel the world and pick up access points, log the geolocation, and upload them to the Wigle Wi-Fi site so that people can search it and see where that access point was found. This is really handy if we were to search for an area and see a particular SSID that was located in Japan, because we can actually use people that have done war driving in Japan, and we can now do the correlation without having to go there. In the same cases, we can add our data and people in other countries or other areas of the world can actually look at the data that we scanned and what access points were picked up and where they were located.

The other tool that we are going to use is a Packet Capture (PCAP) Scanner, which is essentially a Kismet port to the Android. Kismet is a tool that allows for scanning and capturing PCAPs, all the Wi-Fi data into a particular file so that we can go back later and see what was actually being communicated at the access points I attempted. BitShark Share is a paid wireless PCAP tool that is available on

Android; it's about $3.00, and I find it very helpful for when you are actually doing the PCAP scanning, to be able to make sure that it is actually working correctly. Essentially it's Wireshark for Android. So if you were going to use a laptop, I would recommend either using Kismet or Aircrack, both of which work on Linux or Windows fairly seamlessly. If you have a MAC, I would recommend getting a Linux VM or Windows VM and using it on there. In most of my testing, I have not been able to find a tool that works well with MAC so it's easier to purchase a connection between the ALFA cord and the VM and just do it there. That being said, once you've actually captured the PCAP using Aircrack or Kismet or even Android, you can use Wireshark or tcpdump on MAC, Linux, or Windows.

Software Needed - WiGLE Wifi

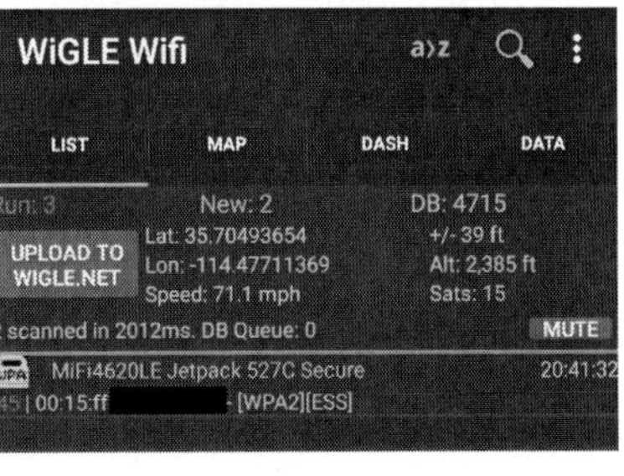

Here is an example of captures that I was doing, with some War Driving with Wigle Wi-Fi. As you can see in the top left, you have a Wigle Wi-Fi icon, and once you click on that you open up a dashboard and it allows us to see latitude, longitude, and speed of the moving device, and altitude, and the SSID—which in this case was a Jetpack MIFI provided by Verizon. The MAC address of that device is visible, and also in the bottom left here, we have the decibel that tells us how close we are to the device and it was −45. It also shows the timing of when we saw this device first, and it allows us to upload that to the web so that we can actually have it for later tracking and provide it to others as well.

Software Needed - WiGLE Wifi

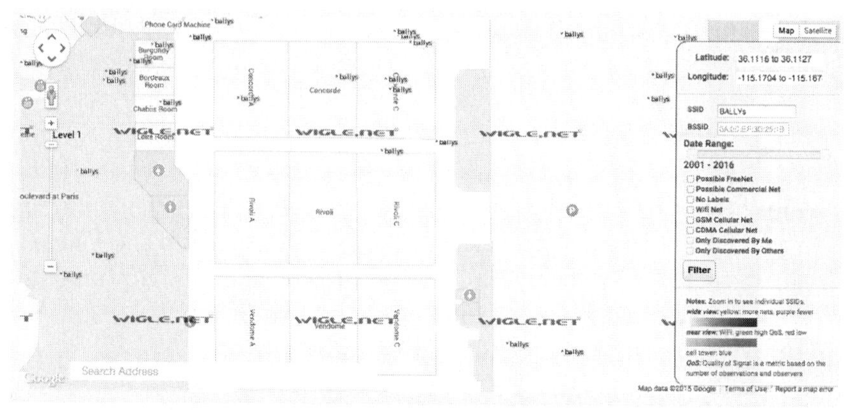

Here is an example of what it looks like after we upload the data. In this particular instance, I was at Blackhat and I was doing some searches for SSIDs representing *BALLYS* Wi-Fi, the hotel hosting the conference. You have latitude and longitude, and you can also see the range of data. Notice that they have actually been doing this for a very, very long time, back from 2001 all the way up to them having the database, to 2016. You can do various filters on it and you can also search by address (see the bottom left of the image), and I'll tell you how that's helpful later on.

Software Needed - PCAP Scanner

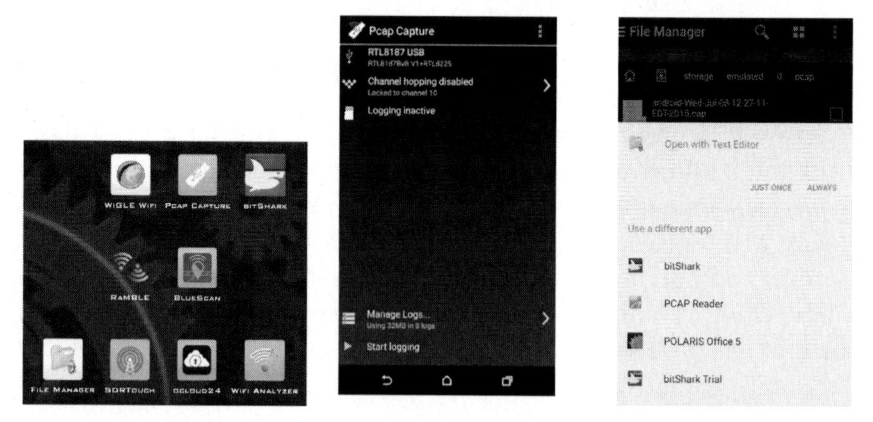

Here is an example of PCAP capture. The icon for that is in the top middle. We have the RTL8187 already connected, and you have the option of selecting a channel so this is a little more helpful if you actually know the SSID and where it's running, but generally I'll leave this to just do what's called channel hopping so we have no restrictions there and we get all the traffic we want. We go ahead and click start logging, and we can also look at previous logs or current running logs in the manage logs section. Over here on the right, this is what the manage log section looks like and we can export that data to any app on the Android that supports PCAPs or text editing.

Software Needed - BitShark Scanner

In order to verify that the files are actually working correctly, here is an example of BitShark Scanner. Notice the icon for that is in the top right; in the middle we have the menu screen so we have eth2, which is the Ethernet port that was connected at that time. We have the time and date as well as MAC address, and we have the size of the actual packets that were captured. This shows number 713, the size of the PCAP, and it shows framed details and Hex information. We can actually see that the PCAP is being communicated successfully, and we are able to capture it as well. It also has multiple filters (although we don't want to necessarily dive into that in this book).

Software Needed - Kismet

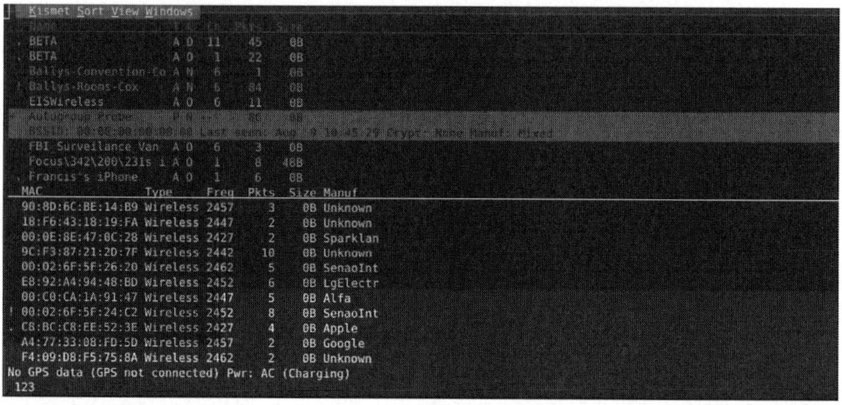

```
Kismet Sort View Windows
 BETA                  A O  11    45    0B
 BETA                  A O   1    22    0B
 Ballys-Convention-Co  A N   6     1    0B
 Ballys-Rooms-Cox      A N   6    84    0B
 E15Wireless           A O   6    11    0B
 Autogroup Probe       P N  ---   80    0B
 FBI Surveillance Van  A O   6     3    0B
 Focus\342\200\231s i  A O   1     8   48B
 Francis's iPhone      A O   1     6    0B
MAC               Type     Freq Pkts  Size Manuf
90:8D:6C:BE:14:B9 Wireless 2457    3    0B Unknown
18:F6:43:18:19:FA Wireless 2447    2    0B Unknown
00:0E:8E:47:8C:28 Wireless 2427    2    0B Sparklan
9C:F3:87:21:2D:7F Wireless 2442   10    0B Unknown
00:02:6F:5F:26:20 Wireless 2462    5    0B SenaoInt
E8:92:A4:94:48:BD Wireless 2452    6    0B LgElectr
00:C0:CA:1A:91:47 Wireless 2447    5    0B Alfa
00:02:6F:5F:24:C2 Wireless 2452    8    0B SenaoInt
C8:BC:C8:EE:52:3E Wireless 2427    4    0B Apple
A4:77:33:08:FD:5D Wireless 2457    2    0B Google
F4:09:D8:F5:75:8A Wireless 2462    2    0B Unknown
No GPS data (GPS not connected) Pwr: AC (Charging)
123
```

If you are going to actually do a capture on a laptop, I would recommend only doing that when you are in an environment that condones laptop. So if you were at a coffee shop or you work where there are hotspots where people are also working on laptops, only do it there. If you are out in the mall, you don't want to carry around a laptop because that might look creepy. In this situation, we have selected the auto probe group and essentially this shows us several different devices that are attempting to connect to SSIDs but have not successfully connected. We have an Apple device, we have an Android, and someone was using an ALFA cord as well. We can parse through this, and it's all being recorded in a PCAP that we can later go back through and parse.

Software Needed - Wireshark/tcpdump

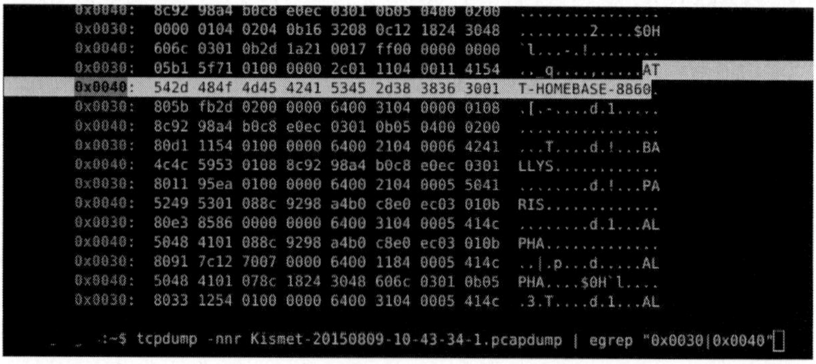

```
0x0040:  8c92 98a4 b0c8 e0ec 0301 0b05 0400 0200  ................
0x0030:  0000 0104 0204 0b16 3208 0c12 1824 3048  ........2....$0H
0x0040:  606c 0301 0b2d 1a21 0017 ff00 0000 0000  `l...-.!........
0x0030:  05b1 5f71 0100 0000 2c01 1104 0011 4154  .._q.........AT
0x0040:  542d 484f 4d45 4241 5345 2d38 3836 3001  T-HOMEBASE-8860.
0x0030:  805b fb2d 0200 0000 6400 3104 0000 0108  .[.-....d.1.....
0x0040:  8c92 98a4 b0c8 e0ec 0301 0b05 0400 0200  ................
0x0030:  80d1 1154 0100 0000 6400 2104 0006 4241  ...T....d.!...BA
0x0040:  4c4c 5953 0108 8c92 98a4 b0c8 e0ec 0301  LLYS............
0x0030:  8011 95ea 0100 0000 6400 2104 0005 5041  ........d.!...PA
0x0040:  5249 5301 088c 9298 a4b0 c8e0 ec03 010b  RIS.............
0x0030:  80e3 8586 0000 0000 6400 3104 0005 414c  ........d.1...AL
0x0040:  5048 4101 088c 9298 a4b0 c8e0 ec03 010b  PHA.............
0x0030:  8091 7c12 7007 0000 6400 1104 0005 414c  ..|.p...d.....AL
0x0040:  5048 4101 078c 1824 3048 606c 0301 0b05  PHA....$0H`l....
0x0030:  8033 1254 0100 0000 6400 3104 0005 414c  .3.T....d.1...AL

:~$ tcpdump -nnr Kismet-20150809-10-43-34-1.pcapdump | egrep "0x0030|0x0040"
```

Here is an example of tcpdump with the variable "-nnr", which is network neighbors on the Kismet PCAP that was exported, as well as using the grep command to parse out a particular hex position in the file of 0x0030 and 0x0040 which contain the SSID of the device that was trying to connect. It was a particular device trying to connect to *AT&T homebase 8860*. Other SSIDs that show up in the PCAP capture were *BALLYS*, and *ALFA*.

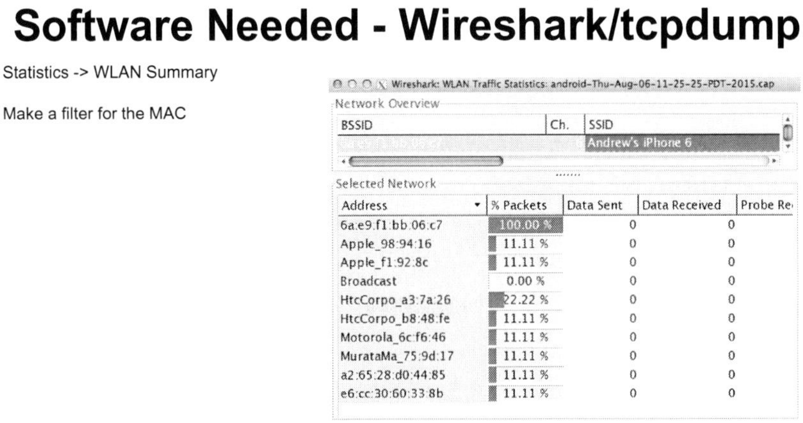

Software Needed - Wireshark/tcpdump

Statistics -> WLAN Summary

Make a filter for the MAC

Wireshark: WLAN Traffic Statistics: android-Thu-Aug-06-11-25-25-PDT-2015.cap

Network Overview

BSSID	Ch.	SSID
		Andrew's iPhone 6

Selected Network

Address	% Packets	Data Sent	Data Received	Probe Re
6a:e9:f1:bb:06:c7	100.00 %	0	0	
Apple_98:94:16	11.11 %	0	0	
Apple_f1:92:8c	11.11 %	0	0	
Broadcast	0.00 %	0	0	
HtcCorpo_a3:7a:26	22.22 %	0	0	
HtcCorpo_b8:48:fe	11.11 %	0	0	
Motorola_6c:f6:46	11.11 %	0	0	
MurataMa_75:9d:17	11.11 %	0	0	
a2:65:28:d0:44:85	11.11 %	0	0	
e6:cc:30:60:33:8b	11.11 %	0	0	

Using the command line to parse PCAPs can be kind of messy, but it is important to understand how to parse things out and use it on the command line so if there is an error or missing data you can adjust for it. If you want to use the GUI—Graphic User Interface—Wireshark is a good solution. In order to get the same information and parse it, from the drop down menu click "Statistics" and then "WLAN Summary". That will go through and take a few minutes to parse, but it will show all the SSIDs that other devices are attempting to connect to. In our sample PCAP, someone was trying to connect to *Andrew's iPhone 6*. This most likely was a tethered iPhone someone was using for Internet. So this shows a whole bunch of devices in the list here, but we can actually make a filter for this MAC that was attempting to go to *Andrew's iPhone 6* and find out what else it was attempting to connect to.

Software Needed - Wireshark/tcpdump

Statistics -> WLAN Summary

Make a filter for the MAC

Filter: wlan.addr==94:94:26:98:94:16 Expression... Clear

```
ength | Info
281 Probe Response, SN=1931, FN=0, Flags=....R..., BI=102, SSID=MPTS
241 Probe Response, SN=1963, FN=0, Flags=........, BI=100, SSID=Andrew's iPhone 6
234 Probe Response, SN=443, FN=0, Flags=....R..., BI=100, SSID=a914809���s
280 Probe Response, SN=1323, FN=0, Flags=....R..., BI=102, SSID=MMD
```

Once we make the filter, we can use “wlan.addr = = XX:XX:XX:XX:XX”, where XX:XX:XX:XX:XX is the MAC address of the associating device. This will present a view showing what other SSID’s the user was trying to connect to. In our sample PCAP, we saw *MPTS* twice, *Andrew’s iPhone 6*, and *MMD*. Now we have all the details that we need in order to start figuring out the attribution of who Andrew really is and where he is from. I’m not going to go into too much detail here because I have some other interesting cases but I would recommend you try this at home with your own devices and see what networks they are trying to connect to. Chances are you will see your home network as well as maybe your work, and a few other locations you have previously visited.

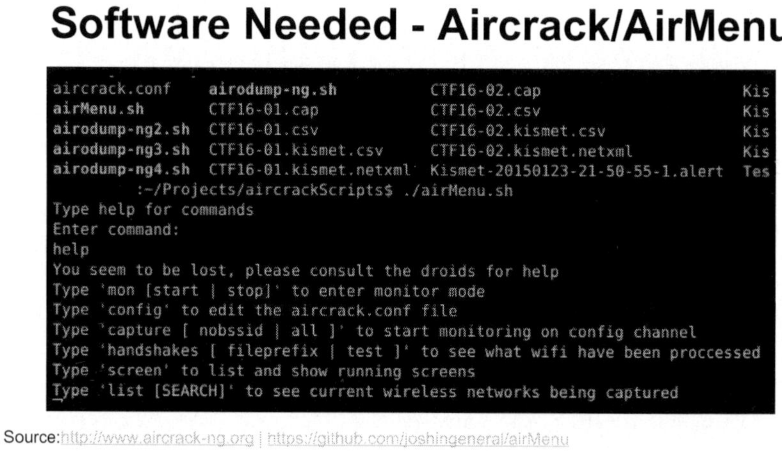

Software Needed - Aircrack/AirMenu

```
aircrack.conf     airodump-ng.sh    CTF16-02.cap                       Kis
airMenu.sh        CTF16-01.cap      CTF16-02.csv                       Kis
airodump-ng2.sh   CTF16-01.csv      CTF16-02.kismet.csv                Kis
airodump-ng3.sh   CTF16-01.kismet.csv    CTF16-02.kismet.netxml        Kis
airodump-ng4.sh   CTF16-01.kismet.netxml  Kismet-20150123-21-50-55-1.alert  Tes
        :~/Projects/aircrackScripts$ ./airMenu.sh
Type help for commands
Enter command:
help
You seem to be lost, please consult the droids for help
Type 'mon [start | stop]' to enter monitor mode
Type 'config' to edit the aircrack.conf file
Type 'capture [ nobssid | all ]' to start monitoring on config channel
Type 'handshakes [ fileprefix | test ]' to see what wifi have been proccessed
Type 'screen' to list and show running screens
Type 'list [SEARCH]' to see current wireless networks being captured
```

Source:http://www.aircrack-ng.org | https://github.com/joshingeneral/airMenu

In order to capture the wireless PCAPs on a laptop we can use a tool Aircrack. Aircrack generally is used for cracking into web or WPA networks; however, it does capture PCAPs so it can be used to actually parse the information out and it has a slightly different interface than Kismet, which some of you might find a little easier to understand.

The first script that I wrote was one that does a switching of the cord from regular mode into monitor mode so we could actually capture packets. It has a configuration file called Aircrack.com, and you can add in certain BSSID or channels in there in order to restrict things down. You also have the ability to capture all SSIDs or make restrictions here. You can also see the number of handshakes, so this is really only useful if you are doing cracking, but as I mentioned, this is an entire menu of all the features for Aircrack.

You also have the "screen" command so you can do multiple scans. Generally, if I am on a laptop, I would use a USB3.0 hub, I'll plug multiple cords into that and have multiple scans going at once so that I can maximize the amount of data I'm getting and not miss packets, because when you are doing channel hopping you have a tendency to miss something. If you are looking at channel 1 while something is happening at channel 6, you'll miss it. The more cords you have, the better chance you have of that not happening.

I'm going to finally list, and you can see all the screens that are available and look at the status update on that.

Software Needed - Aircrack/AirMenu

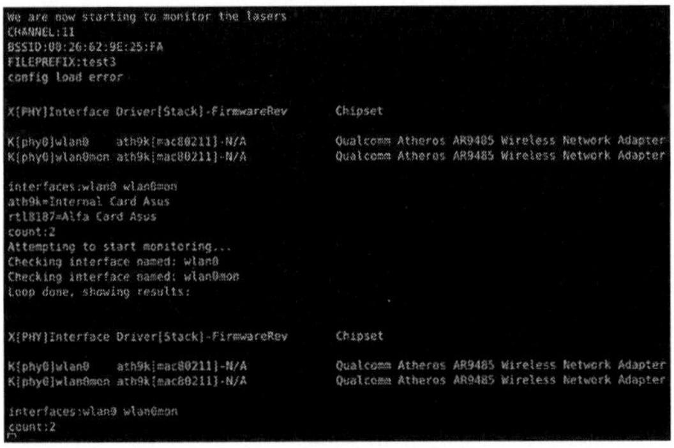

```
We are now starting to monitor the lasers
CHANNEL:11
BSSID:00:26:62:9E:25:FA
FILEPREFIX:test3
config load error

X[PHY]Interface Driver[Stack]-FirmwareRev       Chipset

K[phy0]wlan0    ath9k[mac80211]-N/A             Qualcomm Atheros AR9485 Wireless Network Adapter
K[phy0]wlan0mon ath9k[mac80211]-N/A             Qualcomm Atheros AR9485 Wireless Network Adapter

interfaces:wlan0 wlan0mon
ath9k=Internal Card Asus
rtl8187=Alfa Card Asus
count:2
Attempting to start monitoring...
Checking interface named: wlan0
Checking interface named: wlan0mon
Loop done, showing results:

X[PHY]Interface Driver[Stack]-FirmwareRev       Chipset

K[phy0]wlan0    ath9k[mac80211]-N/A             Qualcomm Atheros AR9485 Wireless Network Adapter
K[phy0]wlan0mon ath9k[mac80211]-N/A             Qualcomm Atheros AR9485 Wireless Network Adapter

interfaces:wlan0 wlan0mon
count:2
```

Source:http://www.aircrack-ng.org | https://github.com/joshingeneral/airMenu

Here is an output of what it looks like when we run the menu. AirMenu is essentially just one menu (Schroeder, 2015). You do certain commands and you select the different pieces from that. Here in this particular one, we had channel 11 at a specific BSSID. I selected a prefix for all the PCAPs to have as test 3. I had already set the interface from WLAN to zero to monitor mode, so you can see the two interfaces there that I have to interact with and here are some more details about the actual cord itself, the chip set, and various other things.

Software Needed - PCAP Scanner

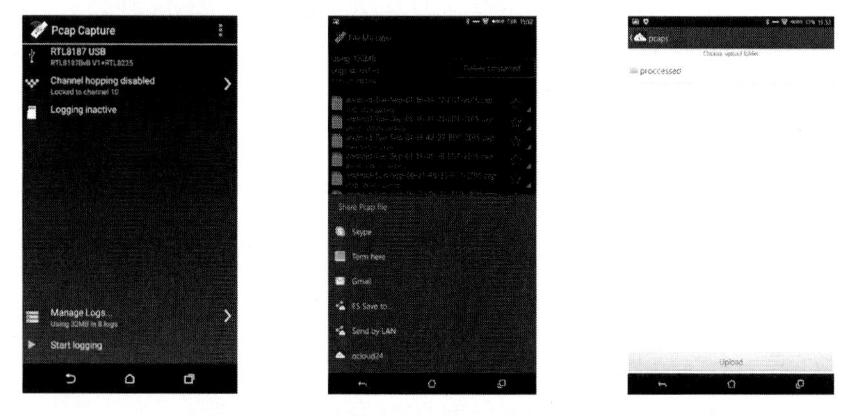

I also created a tool on the Android that would do the backend processing and essentially using T-Shark and tcpdump in order to figure out who is who and get through some more data right on the Android device so you are not having to do too many steps. This is a PCAP scanner, and we have our ALFA cord hooked up via the OTG. We've started logging and we've got several packets here, and we found one that we want to go ahead and send to be processed. In this particular case, instead of sending as an email or opening a text item, I'm just going to share the Packet Capture (PCAP) with OwnCloud. OwnCloud I have mounted on a Linux VM, and that will essentially allow me to run this particular script.

Software Needed - Bash Script

```
No files found
Checking again in:
5
4
3
2
1
Using pcap: /mnt/webdav2/pcaps/a2.cap
Writing out to: /mnt/webdav2/pcaps/a2.cap.txt
tshark: The file "/mnt/webdav2/pcaps/a2.cap" appears to have been cut short in the middle of a packet.
tshark: The file "/mnt/webdav2/pcaps/a2.cap" appears to have been cut short in the middle of a packet.
  % Total    % Received % Xferd  Average Speed   Time    Time     Time  Current
                                 Dload  Upload   Total   Spent    Left  Speed
100 1647k    0 1647k    0     0   236k      0 --:--:--  0:00:06 --:--:--  307k
@SSID=106F3F73ECAB-1
@SSID=12_LagoonFL
@SSID=1f057815de779adn
@SSID=58:b6:33:10:e3:98 (BSSID) 802.11
@SSID=62_SouthSeasD_3
@SSID=Access Control
@SSID=Alsafir_803
@SSID=Andrew's iPhone 6
@SSID=AP2000
@SSID=Apple Store
@SSID=AS8000
@SSID=AS8001
@SSID=ASUS
@SSID=ASUS_5G_Guest1
```

This script checks every 5 seconds for new PCAPs in a particular PCAP folder, and it parses them out. In this particular one, it says there's an error, it's being cut short; essentially what that means is I was actively scanning at a particular time and didn't close the PCAP out properly. It doesn't mean that you will lose any data; it just means that T-shark is giving a little warning on what's up with that packet.

Here is a listing of all the SSIDs that someone is attempting to connect to, so I quickly list that out and sort so you can see in alphanumeric sorting and find the ones that are of interest to you. So once again here is *Andrew's iPhone 6*.

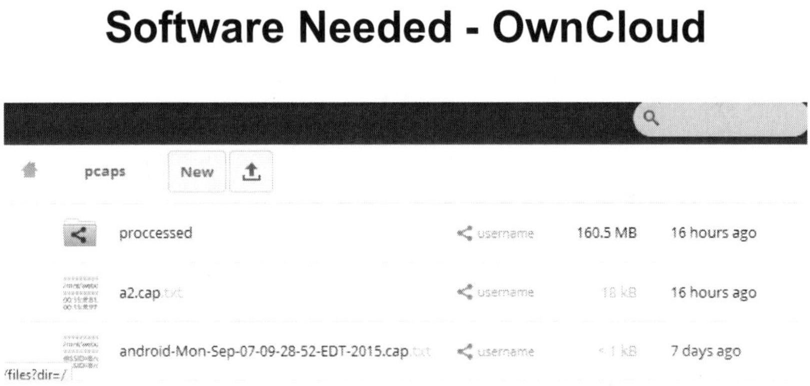

Software Needed - OwnCloud

pcaps New

proccessed	username	160.5 MB	16 hours ago
a2.cap.txt	username	18 kB	16 hours ago
android-Mon-Sep-07-09-28-52-EDT-2015.cap.txt	username	< 1 kB	7 days ago

'files?dir=/

And here is an example of the OwnCloud setup. Notice that we have a process folder and we have two PCAPs here that we've already processed, and any time we have any PCAP in here, essentially it goes through the script and then we move it to the process folder.

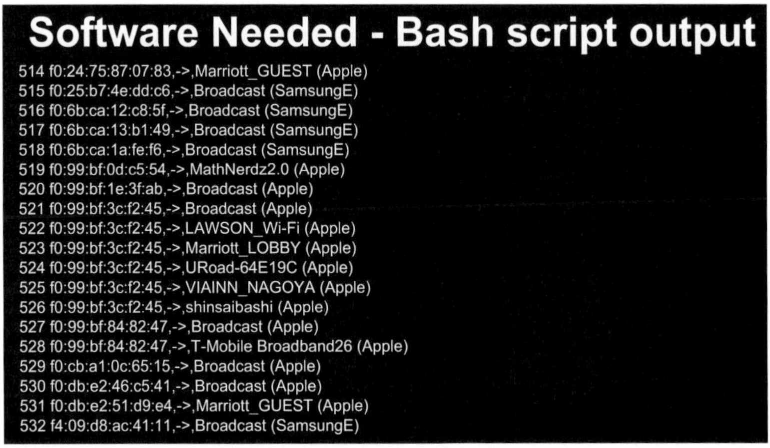

Software Needed - Bash script output

```
514 f0:24:75:87:07:83,->,Marriott_GUEST (Apple)
515 f0:25:b7:4e:dd:c6,->,Broadcast (SamsungE)
516 f0:6b:ca:12:c8:5f,->,Broadcast (SamsungE)
517 f0:6b:ca:13:b1:49,->,Broadcast (SamsungE)
518 f0:6b:ca:1a:fe:f6,->,Broadcast (SamsungE)
519 f0:99:bf:0d:c5:54,->,MathNerdz2.0 (Apple)
520 f0:99:bf:1e:3f:ab,->,Broadcast (Apple)
521 f0:99:bf:3c:f2:45,->,Broadcast (Apple)
522 f0:99:bf:3c:f2:45,->,LAWSON_Wi-Fi (Apple)
523 f0:99:bf:3c:f2:45,->,Marriott_LOBBY (Apple)
524 f0:99:bf:3c:f2:45,->,URoad-64E19C (Apple)
525 f0:99:bf:3c:f2:45,->,VIAINN_NAGOYA (Apple)
526 f0:99:bf:3c:f2:45,->,shinsaibashi (Apple)
527 f0:99:bf:84:82:47,->,Broadcast (Apple)
528 f0:99:bf:84:82:47,->,T-Mobile Broadband26 (Apple)
529 f0:cb:a1:0c:65:15,->,Broadcast (Apple)
530 f0:db:e2:46:c5:41,->,Broadcast (Apple)
531 f0:db:e2:51:d9:e4,->,Marriott_GUEST (Apple)
532 f4:09:d8:ac:41:11,->,Broadcast (SamsungE)
```

This image shows an example of what the actual output. We have line items here over on the left because I opened it. We have individual MAC addresses that would have been sorted in the second part of the script so that we can actually do correlations here. You'll notice that a lot of these first ones are just not the same and that doesn't really help us too much. We have some broadcasting here which is just merely checking what's around—*Marriott guest, MathNerdz*—individual access devices are actually looking at those. Then, we have this particular device here, which has the same MAC address, and it's connecting to several different points. So this is viable data. We can tell it's an iPhone based on the MAC address; we were able to process that via WireShark, and it has a pretty good database.

Software Needed - Bash Script

1. Defines paths, most of which are mounted on owncloud (No filesize limits)
2. Parses SSID, and MAC's of devices using tshark
3. Builds Header
4. Updates mac address database from wireshark
5. Matches mac addresses with that database
6. Prints all output as SSID, devices mac address association, and vendor type
7. Code: https://github.com/joshingeneral/airMenu

Essentially, first thing it does is it parses the paths, defines them, sees the ones that are mounted, and it parses out SSID, MAC addresses using tshark. It builds a header so we have time and date information, and it updates the MAC address database from WireShark. WireShark keeps track of a whole bunch of different device vendors and tells us what the MAC addresses are so we want to pull that down every time and get it as updated as possible. Then, it does some parsing and it does some matching for our use. Print out all the SSIDs like you saw on the first one—alphanumeric. Then it does the device MAC address association, telling us what MACs the devices are attempting to associate to as SSIDs, and it pipes out the vendor type. And just a quick location of where the script is held; all these scripts are in https://github.com/joshingeneral/airMenu/.

Software Needed - Bash Script

- OwnCloud
- /etc/fstab
  https://192.168.1.2/remote.php/webdav/ /mnt/webdav2 davfs uid=USER 0 0
- Script from https://github.com/joshingeneral/airMenu

```
#!/bin/bash
while true; do
...
#List out the SSID's in order
tshark -nnr $pcap | egrep SSID | egrep "[0-9a-f][0-9a-f]\:[0-9a-f][0-9a-f]\:[0-9a-f][0-9a-f]\:.*" -o | sed 's/802.*SSID\=//g' | awk '{print "@SSID="$4" "$5" "$6} ' | sort -u >> $temp
#List out all the devices attempting to connect
tshark -nnr $pcap | egrep SSID | egrep "[0-9a-f][0-9a-f]\:[0-9a-f][0-9a-f]\:[0-9a-f][0-9a-f]\:.*" -o | sed 's/802.*SSID\=//g' | awk '{print $1","$2","$4" "$5" "$6}' | sort -u >> $temp

#Load mac database from wireshark
curl "https://code.wireshark.org/review/gitweb?p=wireshark.git;a=blob_plain;f=manuf" | awk '{print $1" "$2}' > macs.lst

#Figure out Venders from MAC
while read line;
do
 mac=$(echo $line | egrep "^........" -o);
 vender=$(egrep "$mac" -i macs.lst | awk '{print $2}');
# vender2=$(egrep "$mac" -i macs.lst | sed -e 's/\:/\-/g' | awk '{print $2}');
 if [[ $vender == "" || $mac == "" ]];
  then echo "$line" >> $text
 else
  echo "$line ($vender)" >> $text
 fi;
done < $temp
done
```

To recap, first set up OwnCloud connection; here is the information you need in order to mount that as a web share. Essentially just mount this in your VM and start the script as listed here and let it go through and parse, and it will do a check every 5 seconds at this particular time. You could modify that if you like, but once it finds the PCAP, it will parse it, get out the pertinent information, save it in txt file, and then be ready sitting in your OwnCloud, which you can also have on your Android device to look at, and you can run through the data.

Attribution Wifi - Best Buy & FCPS

```
FC:0A:81:A7:2A:B8 64:C6:67:21:46:60 -76 0-6 3 3
(not associated) BC:20:A4:78:65:FC -52 0-1 0 2
(not associated) 20:7D:74:38:2F:DC -71 0-1 0 12
(not associated) BC:4C:C4:C7:4E:7F -80 0-1 0 1
(not associated) 8C:3A:E3:18:77:54 -81 0-1 0 3
(not associated) 3E:6F:39:43:12:A2 -81 0-1 0 3
(not associated) F0:25:B7:4A:1A:7F -83 0-1 0 4
(not associated) 92:68:C3:03:5C:19 -84 0-1 0 3
(not associated) E8:50:8B:36:BF:31 -84 0-1 0 18 BestBuy,ronali,WPATubez
(not associated) A4:77:33:08:FD:5D -84 0-1 0 24 Verizon SCH-LC11 9f61 Secure
(not associated) 4C:BC:A5:37:40:D5 -85 0-1 0 6
(not associated) 9C:F3:87:55:6C:EF -85 0-1 0 3
(not associated) 00:02:6F:5F:29:00 -87 0-1 0 16 EnGenius
(not associated) 68:09:27:AA:AE:FA -87 0-6 0 8 PARIS
(not associated) 90:8D:6C:BE:14:B9 -87 0-1 0 12
(not associated) 90:B6:86:DE:D0:29 -88 0-1 0 2
(not associated) E0:CB:1D:58:79:3F -88 0-5 0 1
(not associated) FC:0A:81:D9:75:A0 -89 0-6 0 1 smart-rf
(not associated) 00:02:6F:5F:26:20 -89 0-1 0 21 EnGenius
(not associated) C8:AA:21:1A:13:6B -89 0-1 0 10 attwifi,belkin.4b4.guests,ccc-wifi,CoffeeBeanWiFi
(not associated) A4:4E:31:92:98:98 -89 0-1 0 7
(not associated) E0:CB:1D:99:29:4A -89 0-1 0 4 FCPSGuest,HP-Print-92-Photosmart
6520,FCCPublic,CenturyBuilding11A
```

This one is a demonstration of how we can actually do the association. In this particular one, we have captured this on a laptop; this is an expanded version of AirCrack. One SSID is *Best Buy* another

SSID, *WPA tubes*, according to my scanning and looking at various different things is an encrypted connection at Best Buy for employees only. The inference that we are making here is that this is most likely an employee that works at Best Buy because you wouldn't be able to connect to these SSIDs, otherwise if you didn't know the password, unless you broke into it. I guess you can do that as well.

In this first example, I wasn't able to find too much out, but this last one was more interesting. We have *FCPSGUEST*, which is connected to a printer; we also have *FCCPublic* and finally, *Century Building 11A*. So let's see what I was able to pull out.

Attribution Wifi - FairFax County

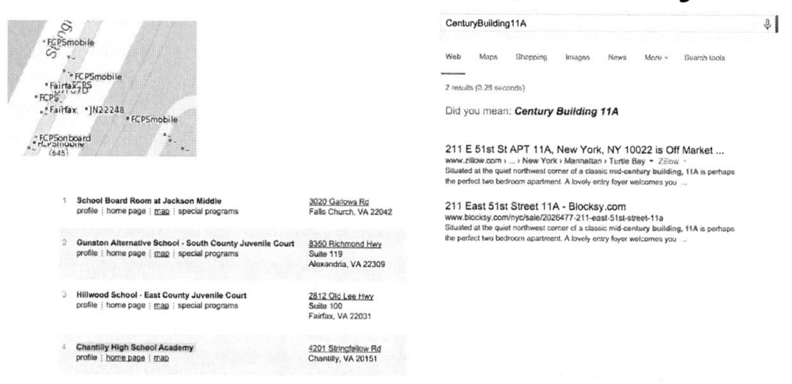

With Wigle Wi-Fi, I was able to find that the *FCPS mobile* was the syntax used by the Fairfax County's school system. Even though I don't actually see the individual SSID here, this could be because they set it up after someone did the scan. The syntax is fairly solid. The other thing to note is that it's possible, because it says *FCPS mobile* here and on board that since this person was connecting to *FCPSGuest*, they were just simply visiting one of the Fairfax high schools and were not actually connected to an employee or student or teacher network that would have been protected.

The other thing that I want to note is that when I looked up information for *Century Building 11A*, I found a few things in New York, but I wasn't able to find a solid SSID. That could be one thing that's a little bit difficult with association and one thing just to keep in mind is even though this is a really good technology for finding people, sometimes you will hit a wall and you won't be able to find anything. So that's just something to keep in mind.

Attribution Wifi - FCC

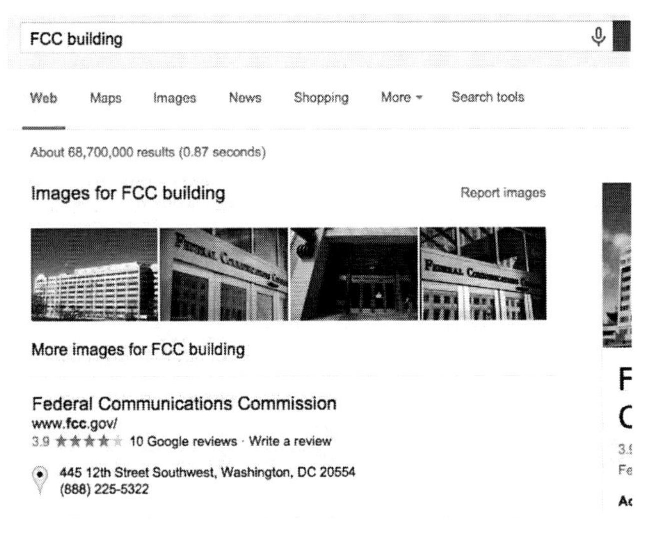

Attribution Wifi - FCC

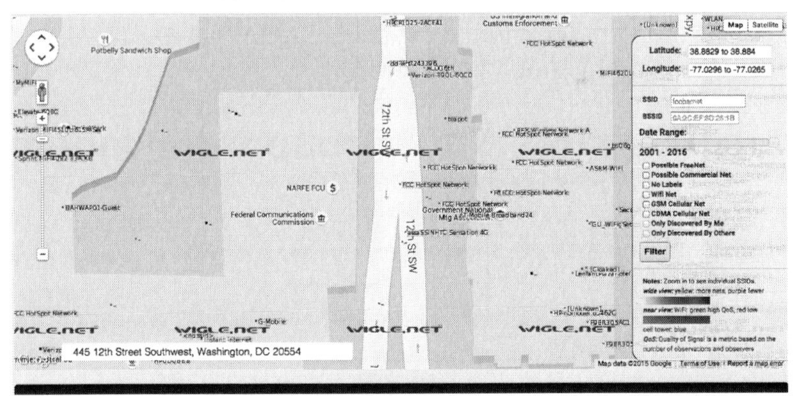

For the FCC Building, the approach I took was to google the FCC Building and then type that address into Wigle Wi-Fi and look for that SSID in that general area. For the FCC Hotspot Network, that's pretty much all that I see around here. As I mentioned earlier, it doesn't always necessarily work. In this particular instance, we hit a wall and do not see an SSID. Once again, this could be because they set up that network after someone did a war drive, so we could go over there and run a Wigle Wi-Fi and see if we could find that access point, but we already have a little bit of attribution and we can make some inferences based on the SSID name.

The other one that was really helpful was finding a group that had gotten off a tour bus at a particular hotel that I found of interest to do some scanning on. There were some people that got off the tour bus and I just hung out in the lobby and scanned while they checked in. Out of that scanning, I was able to find out that there was an Apple device that had connected to these several networks.

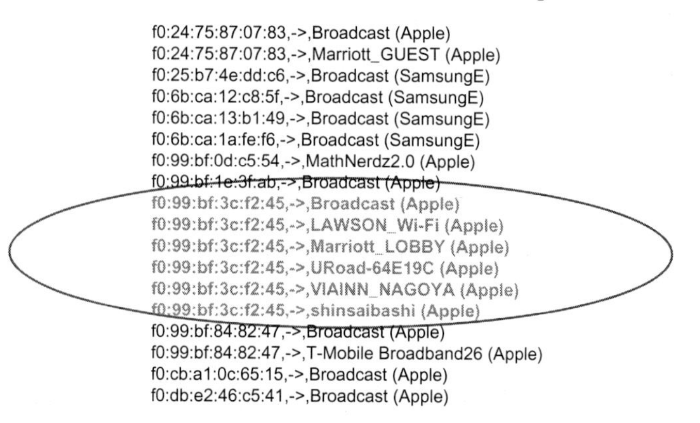

Some of these do look like they are Japanese, but I wanted to find out their exact origination.

I first googled the name of this first hotspot and found it to be a central area in Japan, so that was pretty solid. The second thing I did was I googled the "Via Inn Nagoya," and I found that it was in the western portion of Japan. And then finally I looked around and had to do a little bit more deep digging and found that there was a Lawson ID Japanese printing station that had also been connected to by that particular device. From that information, I can determine that those individuals had most likely come from Japan.

Moving on to the next piece of information that we can find out about individuals, which is through Bluetooth. There are several tools that we can use: Ramble, on Android; BlueScan, which is also on Android; and if you have an iPhone, you can use Bluetooth Smart Scan to figure out information on there. But in this book, we will be focusing mainly on the Android and some scripts that I have written for that to help us figure out information.

Software Needed - Ramble

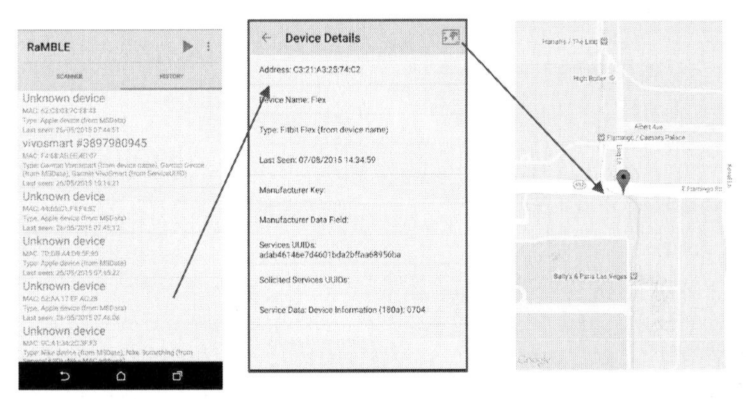

Ramble is a solid tool, and it gives a lot of information. From the first screen here where you do the scanning, I can do both a low energy scan and a classic scan. It shows lots of data, so you can see here this was an Apple device, and you have a Nike device down here, and a Garmin device as well. If you click on each of the individual devices, even though it says unknown device, you can go through and actually figure out key pieces of information.

This one was one that was a little further down the list. It identifies itself as a Flexbit, which is a Fitbit device, which is normally contained within the rest of the individual for tracking health information. And finally, it also keeps the coordinates of where the device was seen and maps that out right on the Android device. I really liked Ramble and started to use that quite a bit; however, I did see some limitations partly because it exports all the data into a SQLite database and I wanted something I could use with more of an ability to script and do some deeper correlation on.

Software Needed - BlueScan

The other device software piece that I found was called BlueScan. For BlueScan, there isn't as much data, as you'll notice right in the menu screen here where you are actually doing the scanning. However, the information that's exported is all in JSON. The other neat thing about BlueScan is that it does historical data on all the devices that it has seen. For example, this particular device here, I saw it earlier in the morning at 11:30 while I was at a conference and then later on in the day I saw it at 15:13. I was able to look around and make correlations based on those who I had seen in the morning and who I saw later on, and who that individual person was. I also had the details based on the MAC address that it was a Galaxy Samsung device. I can export the data into JSON and I can also do the tracking via the history, and that makes it really valuable to me.

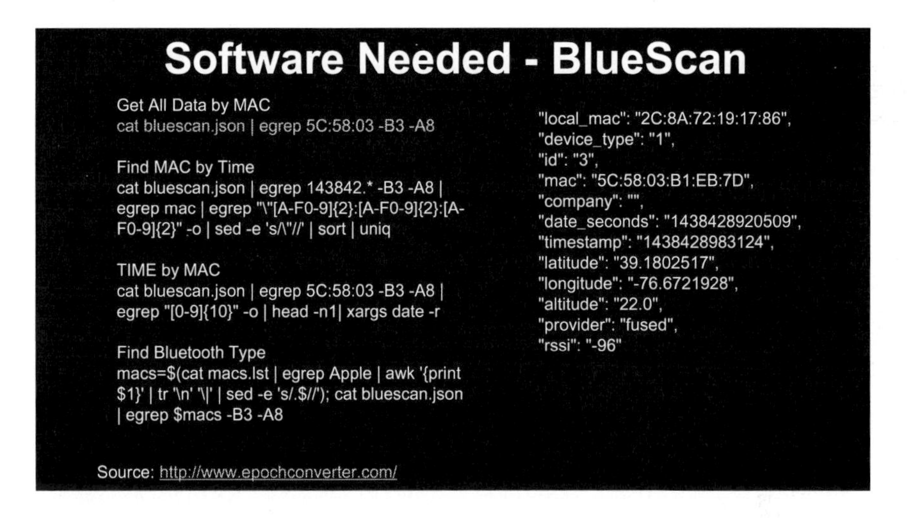

I actually normally use this one a little bit more in my day-to-day scanning, so that's allowed me to create some really cool scripts that allow me to parse some information out and make some correlations.

The first one is once I export to JSON, I'll take it and put it into a text file and I will grep it for all the MAC addresses. This will allow me to then make correlations based on time. The time for the JSON is actually put in Epoch time, which essentially is the number of seconds elapsed since January 1, 1970, not counting leap seconds. What I can do is pick a time within my frame of interest and then I can use the MAC addresses that I have previously gathered in order to get some data from that.

Software Needed - Epochconverter

Get All Data by MAC
cat bluescan.json | egrep 5C:58:03 -B3 -A8

Find MAC by Time
cat bluescan.json | egrep 143842.* -B3 -A8 | egrep mac | egrep "\"[A-F0-9]{2}:[A-F0-9]{2}:[A-F0-9]{2}" -o | sed -e 's/\"//' | sort | uniq

TIME by MAC
cat bluescan.json | egrep 5C:58:03 -B3 -A8 | egrep "[0-9]{10}" -o | head -n1| xargs date -r

Find Bluetooth Type
macs=$(cat macs.lst | egrep Apple | awk '{print $1}' | tr '\n' '\|' | sed -e 's/.$//'); cat bluescan.json | egrep $macs -B3 -A8

The **Unix epoch** (or **Unix time** or **POSIX time** or **Unix timestamp**) is the number of seconds that have elapsed since January 1, 1970 (midnight UTC/GMT), not counting leap seconds (in ISO 8601: 1970-01-01T00:00:00Z).

Source: http://www.epochconverter.com/

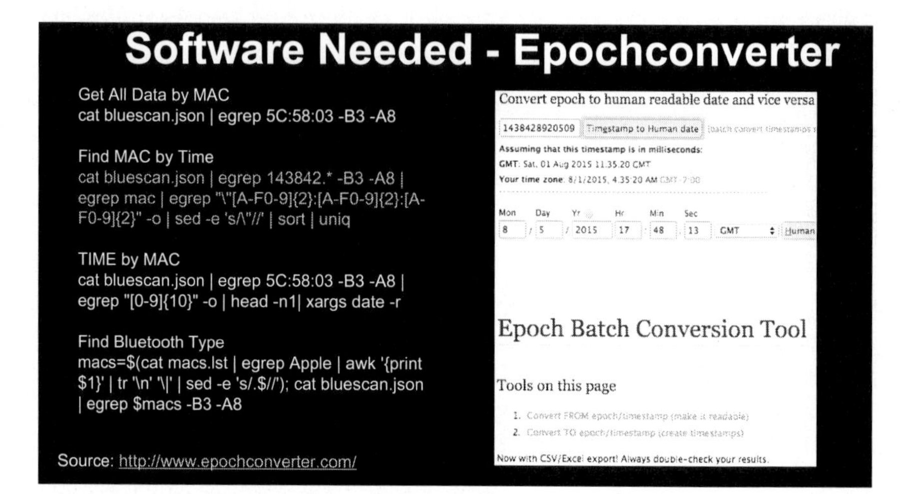

If you want to convert the actual time from epoch scan to human readable, you can use epochconverter.com and then essentially put in the epoch time and it will parse it out in GMT for you. That's how I was able to figure out what time I wanted to put here in the MAC address scan by time filter in order to post that information out.

Software Needed - BlueScan

Get All Data by MAC
cat bluescan.json | egrep 5C:58:03 -B3 -A8

Find MAC by Time
cat bluescan.json | egrep 143842.* -B3 -A8 | egrep mac | egrep "\"[A-F0-9]{2}:[A-F0-9]{2}:[A-F0-9]{2}" -o | sed -e 's/\"//' | sort | uniq

TIME by MAC
cat bluescan.json | egrep 7C:C3:C8 -B3 -A8 | egrep "[0-9]{10}" -o | head -n1| xargs date -r

Find Bluetooth Type
macs=$(cat macs.lst | egrep Apple | awk '{print $1}' | tr '\n' '\|' | sed -e 's/.$//'); cat bluescan.json | egrep $macs -B3 -A8

2C:8A:72
49:A1:FA
54:05:FE
5C:58:03
6F:F7:F4
7C:C3:C8

Source: http://www.epochconverter.com/

Software Needed - WireShark

Get All Data by MAC
cat bluescan.json | egrep 5C:58:03 -B3 -A8

Find MAC by Time
cat bluescan.json | egrep 143842.* -B3 -A8 |
egrep mac | egrep "\"[A-F0-9]{2}:[A-F0-9]{2}:[A-F0-9]{2}" -o | sed -e 's/\"//' | sort | uniq

TIME by MAC
cat bluescan.json | egrep 5C:58:03 -B3 -A8 |
egrep "[0-9]{10}" -o | head -n1| xargs date -r

Find Bluetooth Type
macs=$(cat macs.lst | egrep Apple | awk '{print $1}' | tr '\n' '\|' | sed -e 's/.$//'); cat bluescan.json | egrep $macs -B3 -A8

OUI Lookup Tool

OUI search

Results

Source: https://www.wireshark.org/tools/oui-lookup.html |
https://code.wireshark.org/review/gitweb?p=wireshark.git;a=blob_plain;f=manuf

Here is a sample of the MAC addresses that were parsed out from that time scan, and once I have those MAC addresses I can now upload those into the WireShark search for MAC addresses because they also have a Bluetooth option. You can see from the time frame that I picked an Apple device, and an HTC corporation device, as well as a Texas Instruments device.

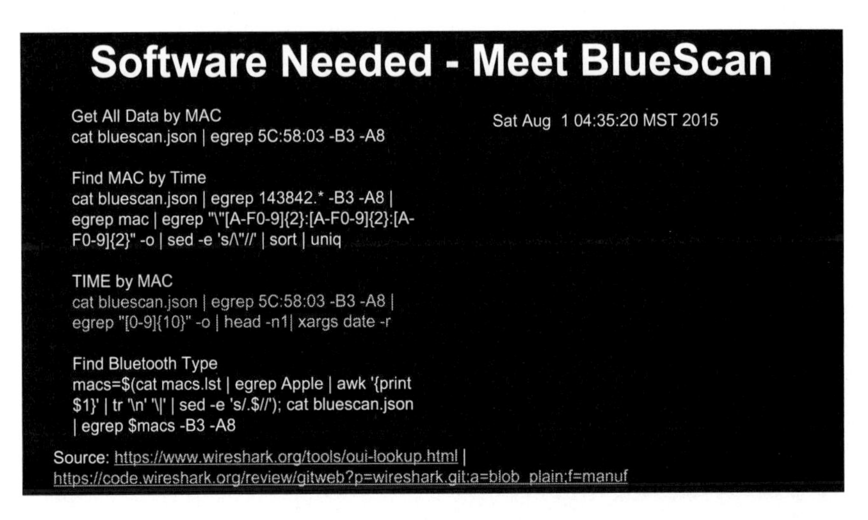

You can also do correlation of the epoch scan right on the command line. If you parse the epoch scan—in this case I grepped one of the MAC addresses I was interested in, I parsed the information with the time stamp and I parsed it with the date-r element—you actually get that on the command line so you could do sorting and figure out some interesting stuff that way as well.

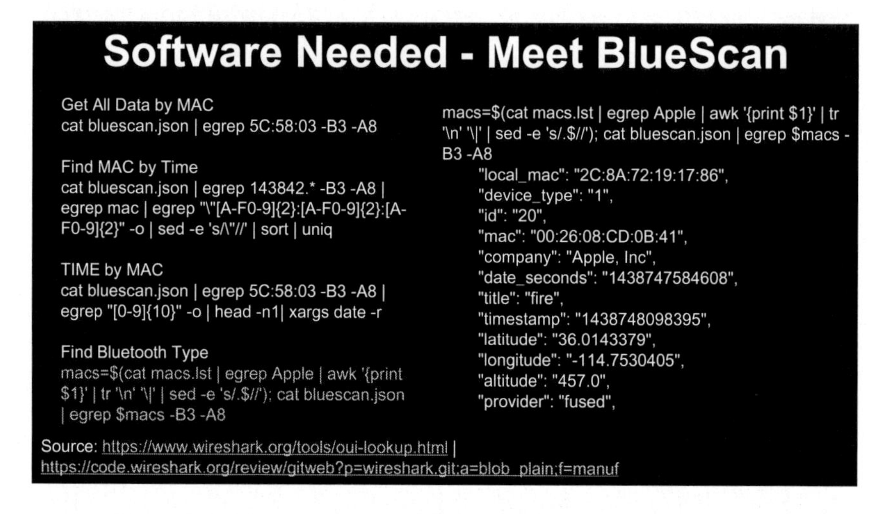

Finally, you can also take the MAC address list that we had from the WireShark, essentially download that into an individual file and you then go through and parse the information and put it right here so you can have a company name that is little bit more detailed and verified. The company name right here is from BlueScan; you are simply parsing out MAC addresses and verifying that this information here is complete from WireShark. And the reason you want to do it from WireShark is because this database is not necessarily updated as frequently as the WireShark database.

Unlike Wigle Wi-Fi, there is no good geo database for Bluetooth, so I created another script that would take the son and put it into csv so we could actually upload it to Google Maps.

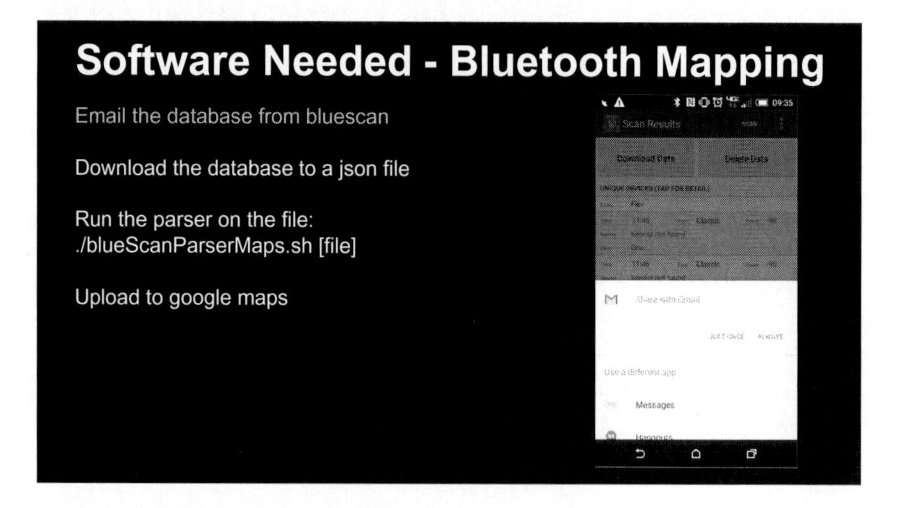

Software Needed - Bluetooth Mapping

Email the database from bluescan

Download the database to a json file

Run the parser on the file:
./blueScanParserMaps.sh [file]

Upload to google maps

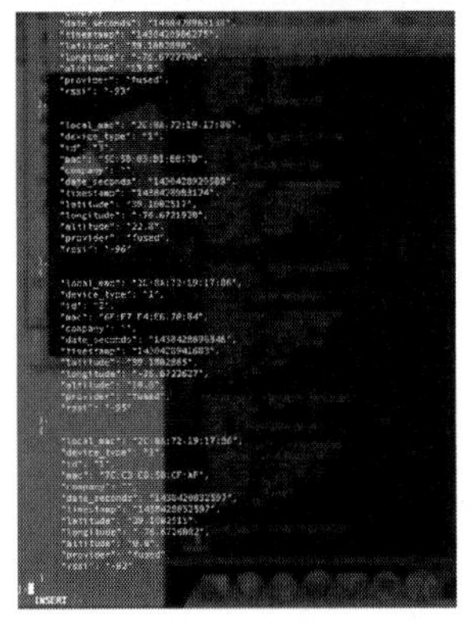

In the app, you can email the JSON file to yourself via Gmail. What you want to do is download that to a JSON file on your local disc and then run the parser on that file.

Software Needed - Bluetooth Mapping

Email the database from bluescan

Download the database to a json file

Run the parser on the file:
./blueScanParserMaps.sh [file]

Upload to google maps:
https://goo.gl/xOJ5sM

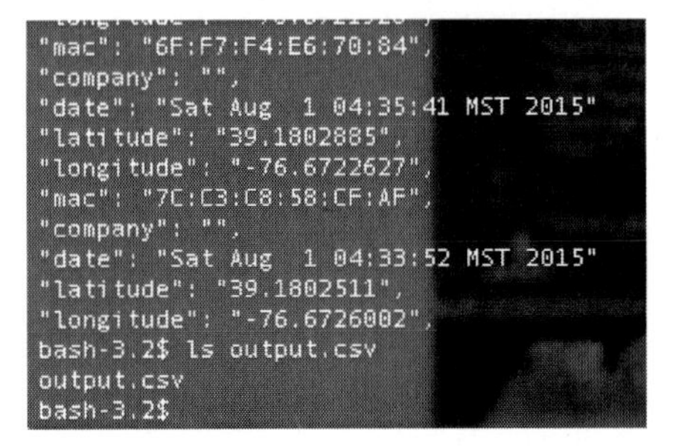

You can see we have latitude, longitude, and MAC address, which we can use in order to build a really cool map.

Software Needed - Bluetooth Mapping

Email the database from bluescan

Download the database to a json file

Run the parser on the file:
./blueScanParserMaps.sh [file]

Upload to google maps
https://goo.gl/xOJ5sM

Software Needed - Bluetooth Mapping

Email the database from bluescan

Download the database to a json file

Run the parser on the file:
./blueScanParserMaps.sh [file]

Upload to google maps

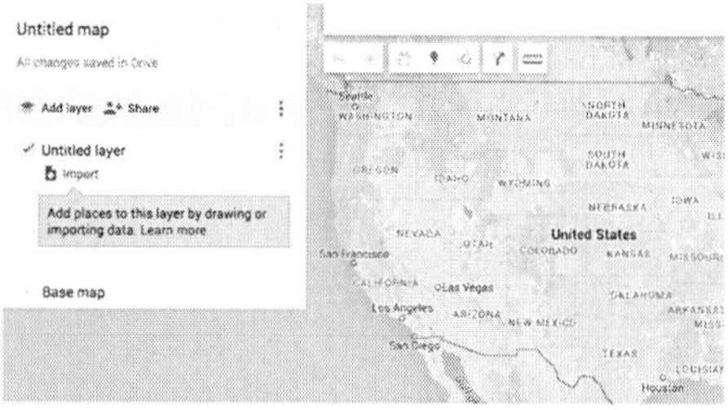

Go into Google Maps, go ahead and click on the search bar, and select my maps. It will bring you to an option to create a map. In that window you have the option to import some extra data so click on that.

Software Needed - Bluetooth Mapping

Email the database from bluescan

Download the database to a json file

Run the parser on the file:
./blueScanParserMaps.sh [file]

Upload to google maps
https://goo.gl/xOJ5sM

Choose a file to import

Upload Google Drive

Drag a CSV, XLSX or KML file here

Cancel

Software Needed - Bluetooth Mapping

Email the database from bluescan

Download the database to a json file

Run the parser on the file:
./blueScanParserMaps.sh [file]

Upload to google maps
https://goo.gl/xOJ5sM

Choose columns to position your placemarks

Select the columns from your file that tell us where to put placemarks on the map, such as addresses or latitude-longitude pairs. All columns will be imported.

Name
Latitude (latitude)
Longitude (longitude)

Continue Back Cancel

Choose a column to title your markers

Pick a column to use as the title for the placemarks, such as the name of the location or person.

Name
Latitude
Longitude

Finish Back Cancel

Edit map title and description

Map title
Defcon

Description
Wifi Bluetooth Data

Save Cancel

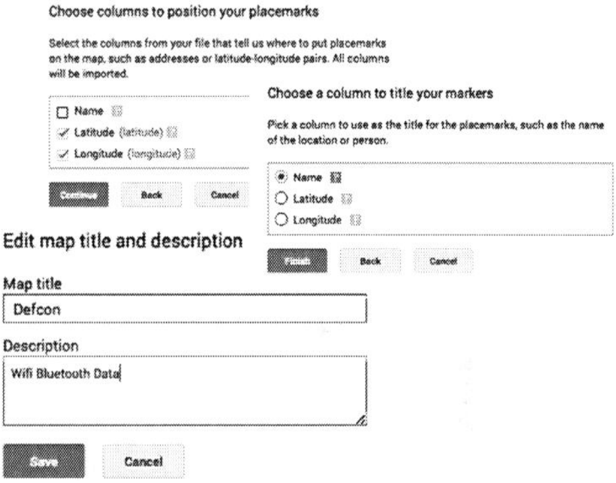

Select the csv file, the bluescanparsermap.sh file, and export it. Select latitude and longitude, which will be really straightforward because of the parsing of the file itself, and when it asks you to select am item for the title of your markers, just select name and click finish. You can also, if you want, give it a map title and a description—I find this useful for going back when I have actually scanned a particular conference and this particular one I did, Defcon.

Software Needed - Bluetooth Mapping

Email the database from bluescan

Download the database to a json file

Run the parser on the file:
./blueScanParserMaps.sh [file]

Upload to google maps
https://goo.gl/xOJ5sM

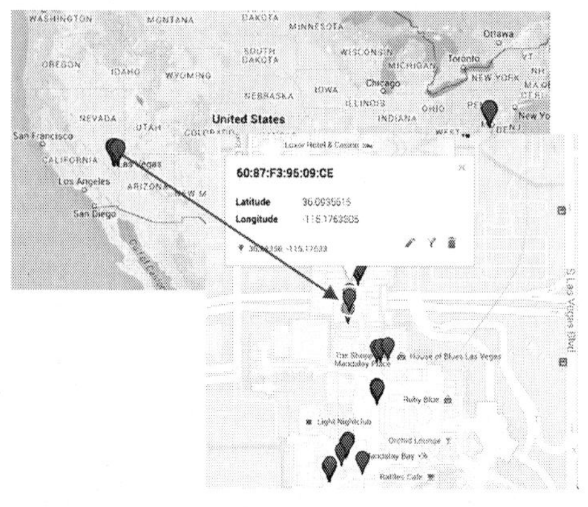

As you can see, once we have done that it now creates a map with all the little pointers and markers that show MAC address, latitude, and longitude. We can also search through this map to see if we have seen a MAC address previously and get points from that as well. In other words, if I saw an individual here in the Maryland or DC area and then I also saw him here in Las Vegas, I confirm that that person moved from here to here via airplane or whatever other means. And that's essentially what I did for this particular search.

Attribution Bluetooth - Girl on a plane

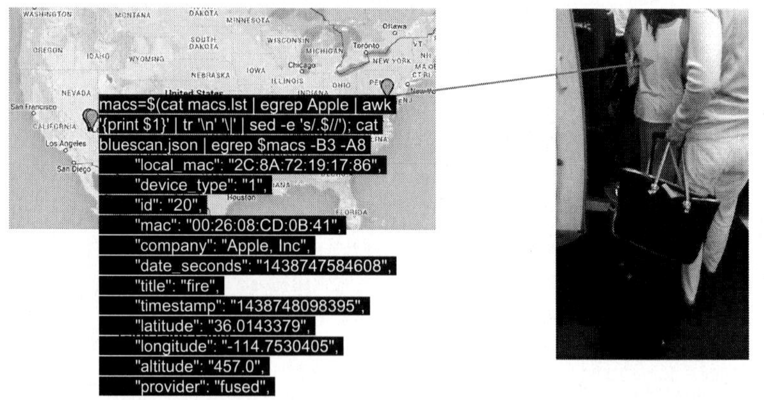

When I was on the airplane in the DC area, I went ahead and ran a scan, which was outputted right here, and I ended up finding this person in Las Vegas as well.

Attribution BT - People in Security Detail

BlueScan SCAN

BLUETOOTH DEVICE DETAILS

GALAXY Gear (8F83)

38:2D:D1:1B:8F:83

SCAN HISTORY

Mode: Classic Scan Thu Aug 06 15:13:17 PDT 2015 Lat/Long: -1.0/-1.0 RSSI: -100
Mode: Classic Scan Thu Aug 06 15:13:05 PDT 2015 Lat/Long: 36.0904337/-115.1749984 RSSI: -97
Mode: Low Energy Scan Thu Aug 06 11:31:58 PDT 2015 Lat/Long: 36.0892714/-115.1781851 RSSI: -90
Mode: Low Energy Scan Thu Aug 06 11:31:58 PDT 2015 Lat/Long: 36.0892714/-115.1781851 RSSI: -87
Mode: Low Energy Scan Thu Aug 06 11:31:50 PDT 2015 Lat/Long: 36.0892714/-115.1781851 RSSI: -88
Mode: Low Energy Scan Thu Aug 06 11:31:44 PDT 2015 Lat/Long: 36.089078/-115.1785294 RSSI: -86
Mode: Low Energy Scan Thu Aug 06 11:31:44 PDT 2015 Lat/Long: 36.089078/-115.1785294 RSSI: -84
Mode: Low Energy Scan Thu Aug 06 11:31:42 PDT 2015 Lat/Long: 36.089078/-115.1785294 RSSI: -84
Mode: Low Energy Scan Thu Aug 06 11:31:27 PDT 2015 Lat/Long: 36.089078/-115.1785294 RSSI: -92
Mode: Low Energy Scan Thu Aug 06 11:31:18 PDT 2015 Lat/Long: 36.089078/-115.1785294 RSSI: -101
Mode: Low Energy Scan Thu Aug 06 11:31:09 PDT 2015 Lat/Long: 36.089078/-115.1785294 RSSI: -93
Mode: Low Energy Scan Thu Aug 06 11:31:01 PDT 2015 Lat/Long: 36.0888924/-115.1785997 RSSI: -87
Mode: Low Energy Scan Thu Aug 06 11:30:51 PDT 2015 Lat/Long: 36.0888924/-115.1785997 RSSI: -93

Here is a little bit of a more detailed description of the information where I was talking about doing the multiple scans. The first scan was 11:31 and then you have a later scan, with the more recent scans being at the top. This can help you if you are not sure or you are in a crowded place—say, on the Metro or on a plane—and you are scanning and you see five different Galaxy S4 phones via the MAC address and you are not sure who is who. You get off the plane and you see three of those people still standing around. Two of them have iPhones and one of them has a Galaxy S4. You could use a scan like this in order to pinpoint and say, "Okay, the person I saw on the phone based on the device that I see in the individual area is this particular individual." Essentially you are using something like Venn diagrams in order to parse that information out.

CONCLUSION

Tips for Protection

- Turn off bluetooth when not needed
- Clean up your wifi
- Be aware of who is around you
- Scan yourself

So what can you do to protect yourself? Turn off Bluetooth when it's not needed and also clean up your Wi-Fi. If you haven't connected to a hotspot or you have changed jobs and are never going to be using that Wi-Fi again, delete it because all that information is essentially giving attackers information about you. Also be aware of those around you. If you see someone doing some kind of weird, suspicious things where they have antenna, you might want to check your location and move out of that area. Finally, scan yourself and become the attacker—see what you are admitting, decide if it's things you feel are acceptable, and that you would be comfortable with someone knowing about you, and if it's not, take actions as I said above in order to remove those items.

REFERENCES

Abraham, J. BlueScan. Accessed October 22, 2015.

Bluetooth SIG, Inc. A Look at the Basics of Bluetooth Technology. 2015. Accessed October 22, 2015 <http://www.bluetooth.com/Pages/Basics.aspx>.

Logitech, Inc. Logitech advanced 2.4 GHz technology. 2009. Logitech.com. Accessed October 20, 2015 <http://www.logitech.com/images/pdf/roem/Logitech_Adv_24_Ghz_Whitepaper_BPG2009.pdf>.

PC Magazine. PC Magazine Encyclopedia. Accessed October 22, 2015 <http://www.pcmag.com/encyclopedia/term/51942/ssid>.

Schroeder, J. Github.com AirMenu—wifi.sh. Accessed October 22, 2015 <https://github.com/joshingeneral/airMenu/blob/master/wifi.sh>.

Schroeder, J. Github.com BlueScanParser. 2015. Accessed October 22, 2015 <https://github.com/joshingeneral/BlueScanParser>.

Shipley, P.M. About Pete Shipley. Accessed October 22, 2015 <http://www.dis.org/shipley/>.

Edwards Brothers Malloy
Thorofare, NJ USA
December 16, 2015